IMAGINATION VACATION
YELLOWSTONE

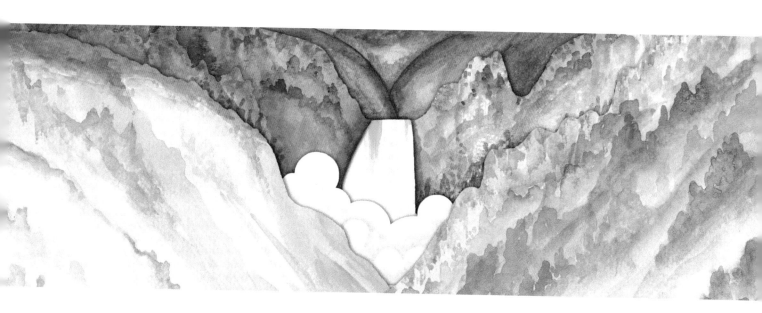

Written by
Anastasia Kierst and Christopher Kierst

Illustrated by
Anastasia Kierst

Many thanks to all of my backers from Kickstarter: Sheri Solomon, Jake King-Gilbert, Angela & Chris White, Beth Pinvises, Andrea & Wyatt Christensen, Janelle & Terry Dunlap, Kristen Elizabeth Rice Jackson, Susan J. Curry, Tyler Holland, Dennis and Megan Cooper, Jondale, Heidi & Carl van Hulst, Claire & Ed Drosdick, Lisa Tuman, Mary Smith, Alexandria & Kevin Trempe, Brian Maloney, Caroline Cochran, Daniel Rhee, Ava Grimm, Brent Cossey, Michelle & Travis Guffey, Claire Trempe, Claudia Meyer, David & Pam Henley, Carla Cooper, Kathleen DiFrancesco, Lindsay Breadon, Mark Bass, Cheryl & Christopher Kierst, Edward Bruce, Luc Arsenault, Jim Ladd, Jamy Beecher, Skip & Barbara Carleton, Franz von Bradsky, Gene Sego

TO MY FAMILY.

Thanks for taking me by the hand on adventures when I was small. Thank you for sharing today's adventures – even when they get a little crazy. Many thanks to my mom, for putting a paintbrush in my hand at such a young age, and to my dad and coauthor for always using the big words with me, even when I was five.

Thanks also to my husband, Jeff Martin, for being supportive and helpful at every turn through my first journey to the land of children's books.

"The joy of youth is discovery."
- Jeff Martin

"Dad?"
"Yes, Emmaline?" Dad replied.
"I think Grand Prismatic is..."

"...a rainbow that curled up for a nap. When it bubbles, the colors escape and color the world. What do you think?" Emmaline asked.

"Well," Dad began, "I think that Grand Prismatic is a *hot spring* – a pool of water that has trickled so deep into the earth that it was super heated by *magma* and..."

"What's magma?" Emmaline interrupted.

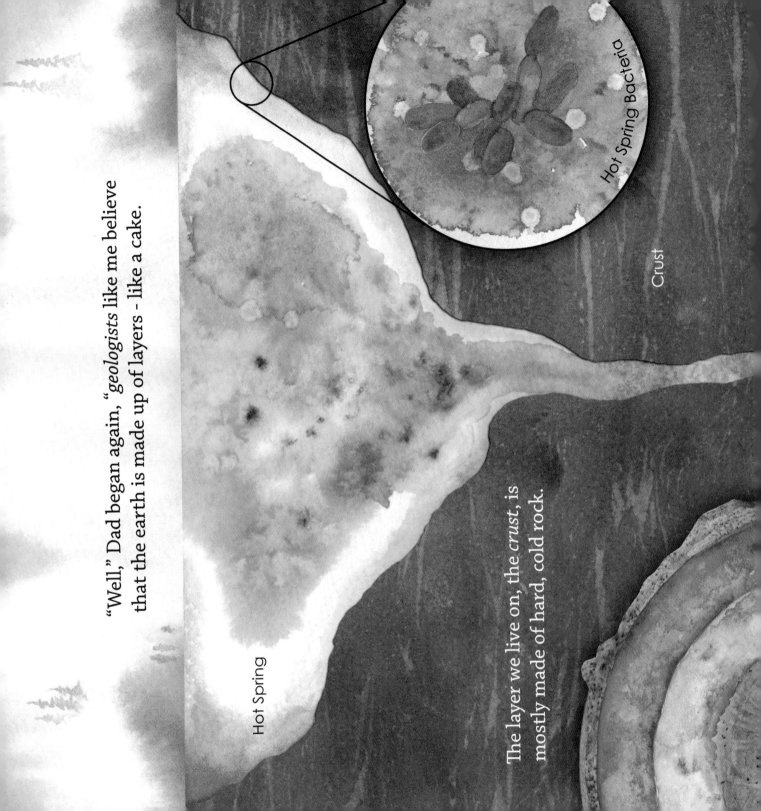

"Well," Dad began again, "*geologists* like me believe that the earth is made up of layers - like a cake.

Hot Spring Bacteria

Crust

Hot Spring

The layer we live on, the *crust*, is mostly made of hard, cold rock.

But the next layer down, the *upper mantle*, is made of very hot, melty rock called *magma*.

The magma heats the water and it goes back up to the surface, like a pot boiling over.

The colors in Grand Prismatic are caused mostly by the algae and bacteria that grow in the spring," Dad explained.

"Really?!" Emmaline asked, raising her eyebrows a bit.

Inner Core

Outer Core

Mantle

Upper Mantle

Crust

"Daddy?"

"What are you imagining, Emmaline?"

"I think Yellowstone Lake is..."

"...an enormous osprey nest. The mamma osprey landed in the park and squished down all the mountains to build the nest. Then it rained and the nest filled with water. What do you think?"

"Well," Dad began, "*volcanologists* think that Yellowstone Lake is part of a huge collapsed *caldera* caused by a *hot spot*."
"A whaty – what? Caused by a *what*?" Emmaline was confused!

Caldera

"Well, a *hot spot* is just what it sounds like. It's a spot on the earth where magma keeps rising.

When the magma stays underground it causes things like the hot spring we saw earlier."

"But when the magma comes out of the ground, it creates volcanoes, like the one we're standing on!"

"What! We're standing on a volcano! Get me out of here!" Emmaline shrieked.

Magma

Hot Spot

"Actually," Dad said,
"we're standing on the edge of an
empty volcano that has collapsed
into a bowl-shape called a *caldera* –
like a cauldron, a witch's pot!"
"Dad, you have a wild imagination!"

"Dad?"
"Yes, Emmaline?"
"I think Castle Geyser is..."

"...a castle built by fairies for the fairy princess to live in," Emmaline imagined. "What do you think?"

"I can see the resemblance," Dad answered, "but I don't know if a fairy princess would want to live there. The geyser shoots out lots of hot water and steam.

Don't you think it would be too noisy and way too wet for fairies? They might get droopy wings.

I think this castle was built by a *geyser*," Dad said. "*Who* built it?" Emmaline asked.

"A *geyser* is kind of like a hot spring," Dad explained, "but the water has to pass through a tight spot. Underground, the water heated by magma is stuck under a layer of cold water. The hot water builds up, until finally it erupts, like that!"

"It's so beautiful!" Emmaline gushed. "But how did it build the castle?"

Geyserite

"Well," replied Dad, "the water carries chemicals. As the water splashes out of the geyser, some of the chemicals are left behind creating a rock called *geyserite*. Every time Castle Geyser erupts, the geyserite castle grows just a little bit more."

Geyser

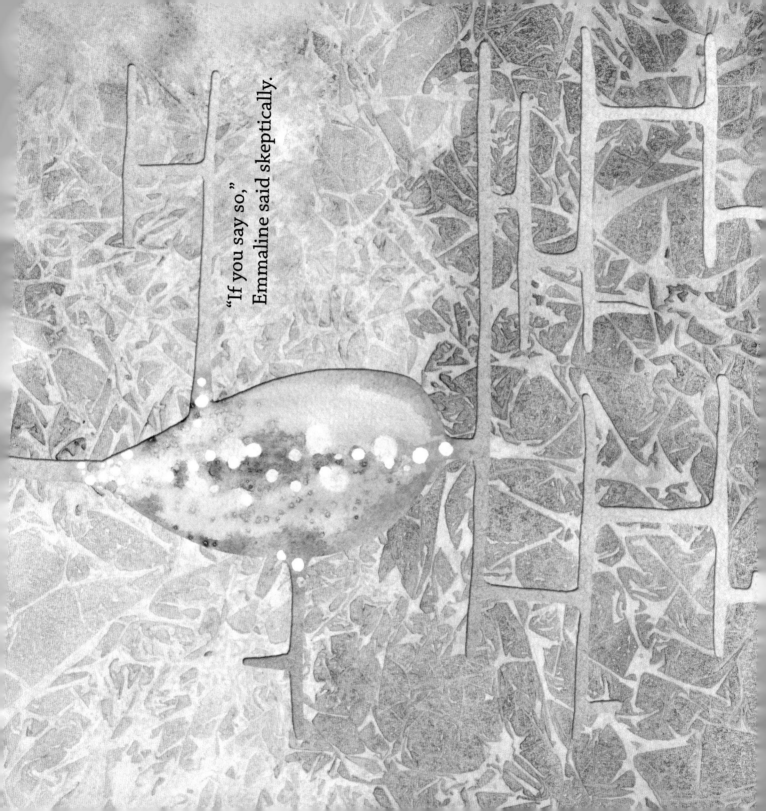

PETRIF
TRE

"...a raven came and sat in it and sang and he was such a TERRIBLE singer that the tree turned to stone so it wouldn't have to listen. What do you think?" Emmaline asked.

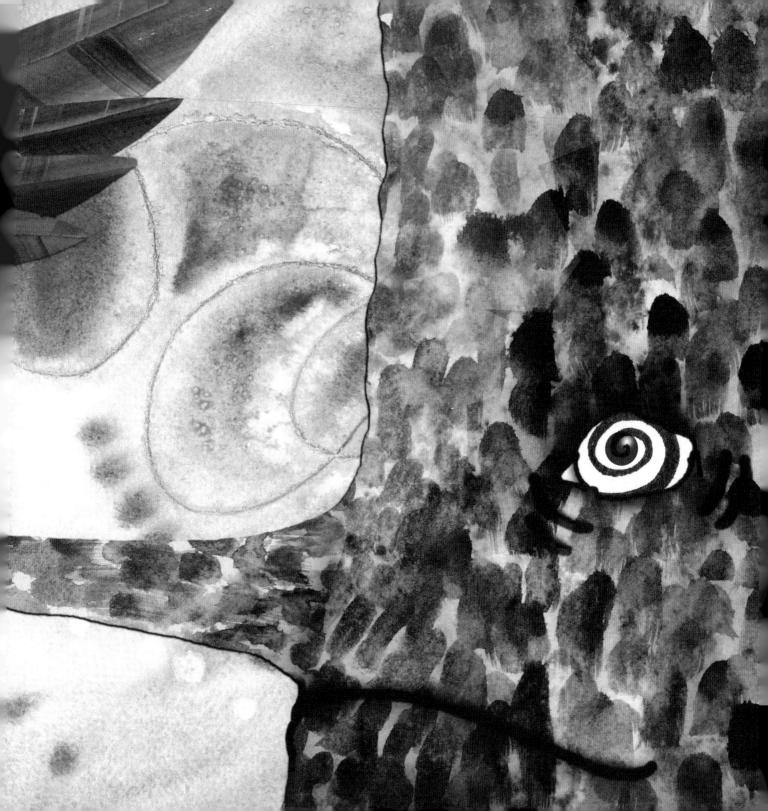

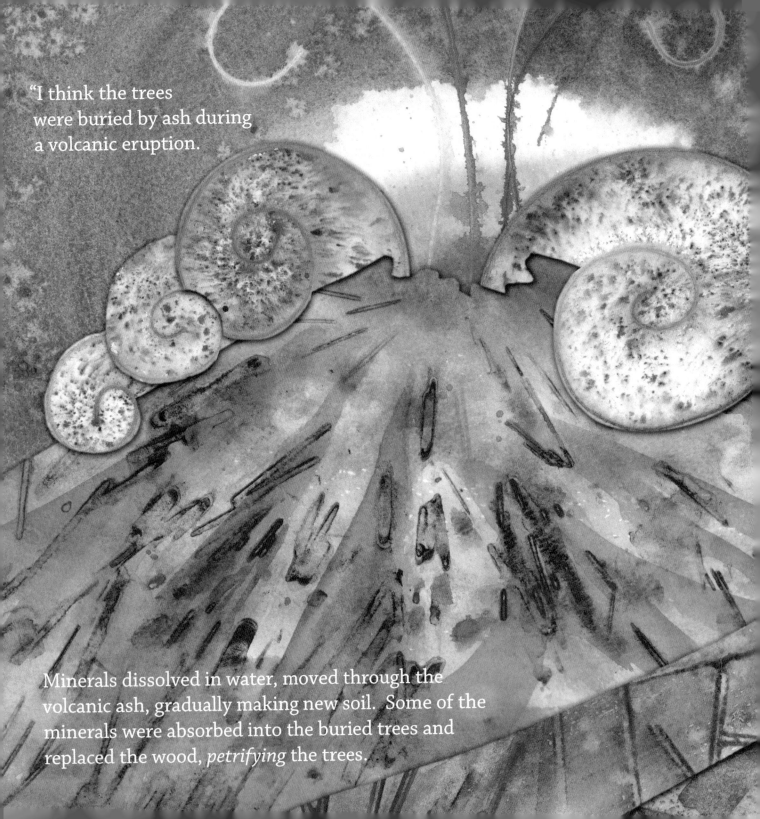

"I think the trees were buried by ash during a volcanic eruption.

Minerals dissolved in water, moved through the volcanic ash, gradually making new soil. Some of the minerals were absorbed into the buried trees and replaced the wood, *petrifying* the trees.

Petrified Trees
in rock layers

Over time," Dad continued, "a new forest grew over the old buried forest. The whole process has been repeated many times here in Yellowstone."
"Yeah, right," Emmaline said, rolling her eyes.

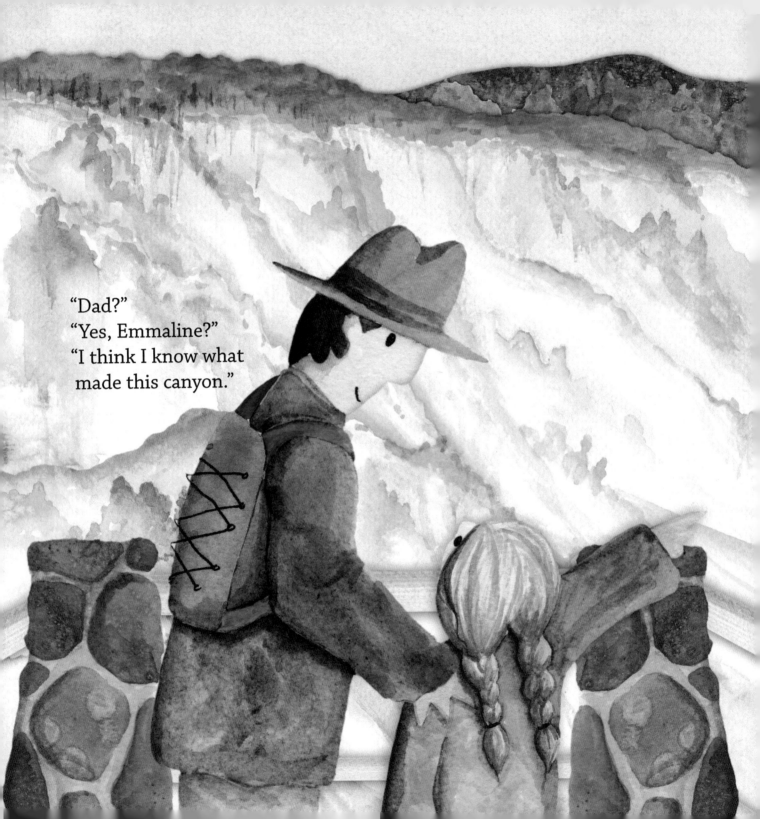

"Dad?"
"Yes, Emmaline?"
"I think I know what
made this canyon."

"I think there was this really big buffalo and a cloud of mosquitoes wouldn't leave him alone and he got so angry he snorted and stomped and scratched this canyon into the ground with his hoof! What do you think?"

"Yeah," Dad agreed, "mosquitoes sure can be annoying. But the Grand Canyon of the Yellowstone is about 900 feet deep and about half a mile wide – way too big for a bison to have made, no matter how angry he was."

Dad went on, "I think your answer is right down there at the bottom of the canyon." "But there's just a river down there," Emmaline argued. "That's right!" Dad said, "It was mostly the river that made the canyon.

Erosion by Water

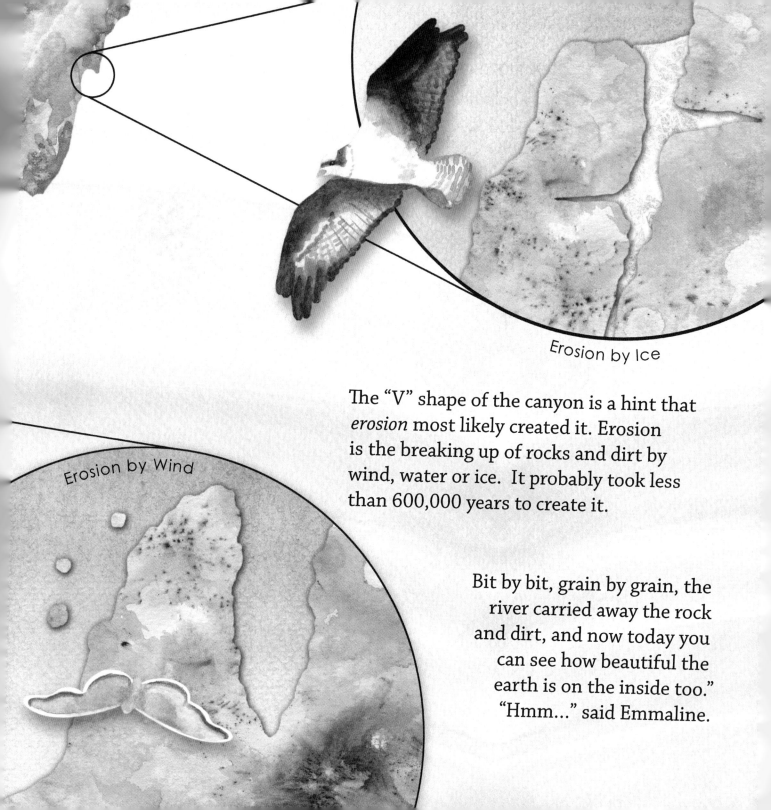

Erosion by Ice

Erosion by Wind

The "V" shape of the canyon is a hint that *erosion* most likely created it. Erosion is the breaking up of rocks and dirt by wind, water or ice. It probably took less than 600,000 years to create it.

Bit by bit, grain by grain, the river carried away the rock and dirt, and now today you can see how beautiful the earth is on the inside too."
"Hmm..." said Emmaline.

"Daddy?"
"Yes, Emmaline?"
"I think Yellowstone is the most magical and amazing and beautiful place in the world! What do you think?"

"I think you're absolutely right!"

GLOSSARY

caldera   An empty volcano that has collapsed.

crust   The outside layer of the earth; the surface we live on.

erosion   The breaking up of rocks and dirt by wind, water or ice.

geologist   A scientist who studies the earth and the forces that change it.

geyser   Similar to a hot spring, but the water has to pass through a tight spot causing it to erupt.

geyserite   The rock made up of chemicals left behind by the water from hot springs and geysers.

hot spring   A hot pool created when water, heated inside the earth, rises to the surface.

hot spot   A spot on the earth where magma rises toward the surface.

magma   The melted rock that makes up the upper mantle layer inside the earth.

petrification   The process by which something that was once alive absorbs and is replaced by minerals.

upper mantle   The layer just underneath the crust inside the earth, made up of melted rock.

volcanologist   A scientist who studies volcanoes.

LEARN MORE

The best way to learn more about Yellowstone is to go there!

Hike, walk the hot spring boardwalks, watch the animals, go to park visitor centers, attend ranger programs and become a Young Scientist or Junior Ranger! When you are old enough, you can even live and work in Yellowstone like the author did.

If you can't visit the park, visit www.nps.gov/yell. You can find some great information on the park including videos and games for kids. Also, try looking for books about Yellowstone at your local library.

20296732R00021

Made in the USA
Charleston, SC
08 July 2013